Friendship at the Feeding Station

Anisha Pokharel
Breanna Epp
Maeve Lausch

Zea Books
Lincoln, Nebraska

Friendship at the Feeding Station
© Copyright 2020 Anisha Pokharel
Illustrations © Copyright 2020 Breanna Epp

ISBN 978-1-60962-169-8

Electronic (pdf) edition available online at https://digitalcommons.unl.edu/zeabook

Print edition available from http://www.lulu.com/spotlight/unllib

UNIVERSITY OF
Nebraska
Lincoln

for Vultures

It's a beautiful day in Pokhara, a city in western Nepal. There are mountains, lakes and beautiful birds.

MONGOLIA

CHINA

NEPAL

INDIA

Winter approaches and so do the birds. A young Steppe eagle and his mother fly to Nepal to escape the harsh winter in their homeland in Mongolia.

'Mom, are we there yet?' asks the weary young eagle.

'Steppe, Look! There is the beautiful Pokhara valley. Let's find something to eat!" replies his mother.

Steppe excitedly sees vultures at a carcass below. 'Let's go there! It looks like a party!'

Mother explains, 'That's a feeding station where a dead cow is left in the field as food for vultures. They don't hunt like us. They eat dead remains. Better to avoid them.'

The eagles have a hard time finding food. They are hungry after the long journey. Farmers poison rats, mice and other small mammals, so each year there is less to eat.

Steppe wants to give the feeding station a try. He lands just far enough away so he can see the carcass. It is quite a sight! So many different kinds of vultures are feasting and fighting for the food. Griffy, a Himalayan Griffon, is the largest. He bullies and tries to chase the others away.

Seeing the fierce, hungry vultures, Steppe is scared. Yet he doesn't give up. He slowly inches towards the carcass and is about to snatch a piece of flesh, when a shadow suddenly falls over him. It is the bully, Griffy!

Griffy chases Steppe down to the ground in anger. 'So, you are stealing my food?'

A terrified Steppe whispers, 'Please let me go. I am sorry. I just needed some food for me and my mother.'

Nearby, Garuda, a beautiful White-rumped vulture, is feeding with his friends. Garuda is kind and friendly, not like Griffy.

Realizing that Griffy was attacking an eagle, Garuda rushes to help.

Angry, Griffy leaves the young eagle. Garuda helps Steppe to clean off his feathers.

'Thank you for saving me. My name is Steppe the eagle.'

Steppe was about to hop away but turns back as Garuda offers a piece of meat to Steppe.

Delighted, Steppe nods to Garuda and flies to share the food with his mother.

The eagles are well fed that day. Steppe is happy with his new friend, Garuda, but he doesn't tell his mother anything.

Next day, Steppe looks for Garuda. Sure enough, Garuda is gliding in the sky. Soon the two new best friends meet every day to play games and explore Pokhara.

One day Steppe's mom finds out about the friendship between Steppe and Garuda.

Steppe comes home from the feeding station. 'Mom, here is food for today.'

Mom is furious. 'Where did you get this food? Have you been hanging out with those vultures? They are mean and filthy. We are eagles, kings of birds. We have a reputation to keep!'

Steppe is surprised. 'But Mom, Garuda helped me. Not all vultures are bad.'

The angry mother tries to calm down. 'You, the majestic Steppe eagle, sought help with vultures? I am worried about you son. They are dangerous, and I don't want you to get injured. Please keep your distance.'

Like every other day, Garuda meets Steppe at their favorite spot at the top of the hill. But he sees that Steppe looks upset: 'Hey Steppe, let's go play!'

Steppe replies, 'I don't want to, I'm sad.'

Steppe explains, 'My mother found out about our friendship and got really angry last night. She doesn't want us to become friends. She said mean things about vultures.'

Garuda sighs, 'I am used to others saying mean things about us, Steppe. Don't be upset. You are my best friend. Follow me. I want to show you something.'

Garuda leads Steppe to the feeding station where other vultures are feeding: 'Look at everyone feeding. There is blood everywhere. What would happen if we had beautiful head feathers?'

Steppe guesses, 'You would get dirty?'

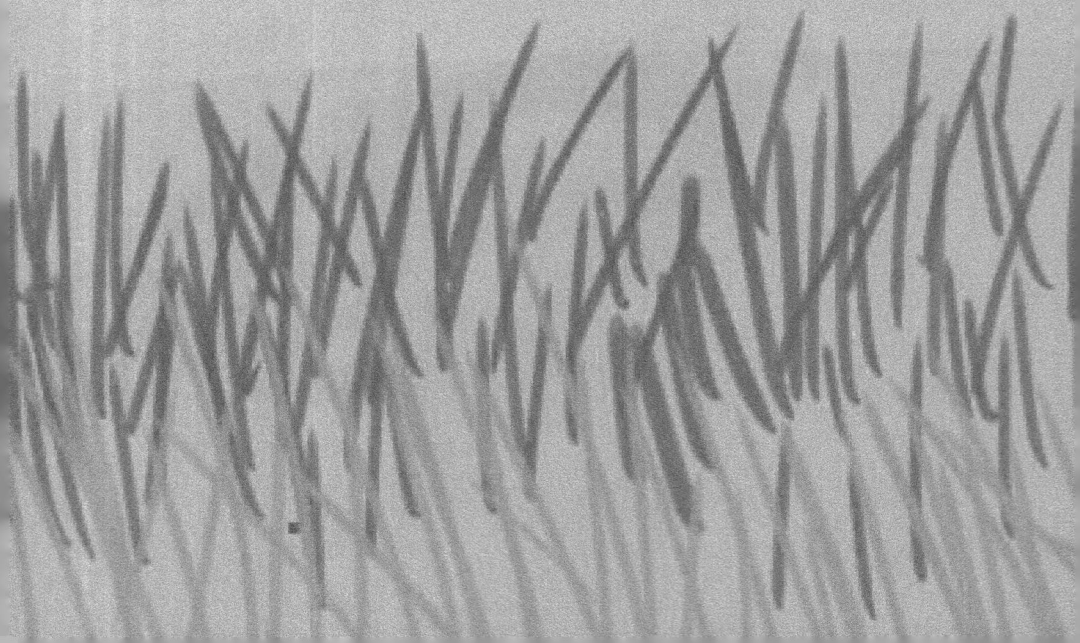

Garuda nods. 'Yes, it would be very hard to get clean, and blood would get stuck. It helps that we have no feathers on our head. We also have very good eyesight and can locate a carcass from far away.'

'So cool. I too have very good eye-sight and a very sharp beak.' Steppe shows off proudly.

Garuda says, 'Steppe, everyone has a role to play in nature. We feed on carcasses. If we were not here, dead animals would pile up and spread diseases. We are nature's cleaning crew. We clean the environment and stop the spread of disease.'

Steppe is impressed. 'We eagles are predators. We hunt other small animals and control their populations, so that they don't destroy the environment.'

Garuda nods. 'You have a job, too. Vultures are disliked because we don't look beautiful. Some years ago, thousands of vultures were killed. It was terrible. Human scientists found that the dead cattle we ate were treated by a painkiller that damaged our kidneys and poisoned us. We almost disappeared, and diseases started to spread to humans.'

Garuda explains, 'Some concerned people created feeding stations to give us safe carcasses. My father says there are more of us now, but vultures are still being poisoned, and our habitat is being destroyed.'

Steppe says, 'My mother said there used to be plenty food before, but now our prey are being killed by people, our forests are being cleared and our lives have become harder.'

Garuda and Steppe say together, 'Life is hard, but we should also have fun. We must be kind to each other. I love you even more, best friend!'

Steppe says, 'Garuda, now I want to take you somewhere'. Together they fly to Steppe's mother.

Steppe introduces Garuda. 'Mother this is my best friend Garuda. He and the other vultures are nature's cleaning crew. They clean carcasses and make sure disease doesn't spread. He also is the kindest bird I know. I really hope you would understand. I learned a lesson not to judge others by the way they look.'

Mother eagle is silent at first.
Then she smiles and says, 'Okay, I
will try to learn from you, Steppe. I
just want you to be safe.'

Since then, Steppe the eagle and
Garuda the vulture remained
best friends.

Vultures of India

Red-headed Vulture
Egyptian Vulture
Eurasian Griffon
White-rumped Vulture
Cinereous Vulture
Slender-billed Vulture
Long-billed Vulture
Lammergeier
Himalayan Griffon

● Endangered/ Critically Endangered

GREEN HUMOUR
www.greenhumour.com

Above: Used with permission from Rohan Chakravarty with Green Humour. 2020

White Rumped Vulture (*Gyps bengalensis*)

Vultures in Asia declined by more than 90% in 1990s due to the anti-inflammatory drug called Diclofenac. It was banned to stop them from going extinct. These vulture restaurants or safe feeding stations ensures the provision of safe food for vultures.

Steppe Eagle (*Aquila nipalensis*)

The Steppe Eagle belongs to the bird of prey family. It preys on small mammals, birds, and carrion. The Steppe Eagle breeds across Romania, Russia, and the Central Asian steppes to Mongolia. While some winter in Africa, others move to South Asia. Many Steppe Eagles have been recorded to migrate through Nepal.

Anisha Pokharel

Anisha Pokharel is a wildlife/raptor biologist from Nepal. She worked with vultures in Nepal and conducted vulture surveys in the USA. She is very passionate about raptor conservation.

About the Authors and Illustrators

Anisha Pokharel is a graduate student at the University of Nebraska–Lincoln studying the diet of migratory behavior of raptors.

Breanna Epp is an undergraduate student studying graphic design. If interested in her illustrations, contact her at breannaepp@gmail.com.

Maeve Lausch is an artist based in Pennsylvania.